Copyright © 2015 by Singapore Math Publishers

All rights reserved.

No part of this book may be reproduced in any form or by any electronic or mechanical means including information storage and retrieval systems, without permission in writing from the author. The only exception is by a reviewer, who may quote short excerpts in a review.

First Edition: September 2015

Introduction

The model diagram is a very important tool in Singapore Math. Students can solve many types of challenging problems with it. The model diagram is actually an algebraic method of using boxes to represent unknown variables. With the model diagram therefore, a student can understand algebraic concepts more visually and concretely.

In this book, we explain how to use model diagrams to solve problems of varying difficulty in a detailed and simple manner. Thirteen basic techniques of solving problems using model diagrams will be taught.

Worked out answers to the problems in this book can be obtained by sending a mail to support@i-ducate.com

This book is suitable for students from grade four to grade six.

Content Page

Chapter 1. Using model diagrams to solve simple problems   3

Chapter 2. Using model diagrams to solve problems involving total   7

Chapter 3. Using model diagrams to solve problems involving one party giving to another   11

Chapter 4. Using model diagrams to solve problems involving one party receiving from another   15

Chapter 5. Using model diagrams to solve problems involving both parties giving

19

Chapter 6. Using model diagrams to solve problems involving both parties receiving

23

Chapter 7. Using model diagrams to solve problems involving one party receiving and one party giving   28

Chapter 8. Using model diagrams to solve problems involving one party giving to the other party   32

Chapter 9. Using model diagrams to solve problems involving insufficient money

36

Chapter 10. Using model diagrams to solve problems involving buying objects that are multiples of another   39

Chapter 11. Using model diagrams to solve problems where the prices of individual objects are unknown   43

Chapter 12. Using model diagrams to solve 'filling up tank' problems   47

Chapter 13. Using model diagrams to solve fraction problems   50

Answers   53

Chapter 1. Using model diagrams to solve simple problems

Problem 1

John has 7 balls. Mark has 5 balls more than John. How many balls does Mark have?

Solution

Draw a box to represent the 7 balls John has and a bigger box to represent that Mark has 5 more balls than John.

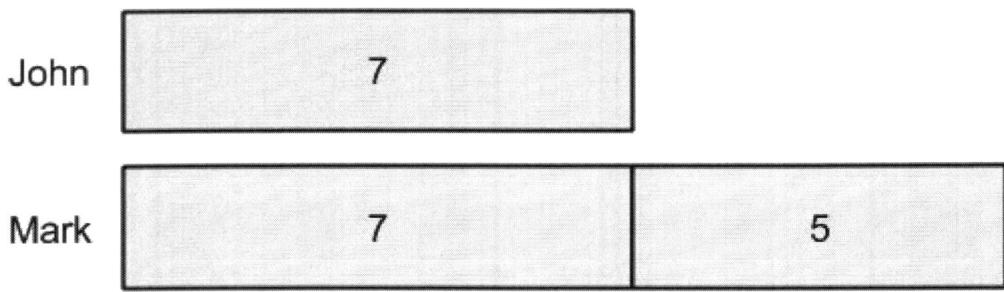

Number of balls Mark has = 7 + 5 = 12

Problem 2

John has 6 balls. Mark has 2 times as many balls as John. How many balls does Mark have?

Solution

Draw a box to represent the 6 balls John has and another two boxes of the same size to represent the number of balls Mark has.

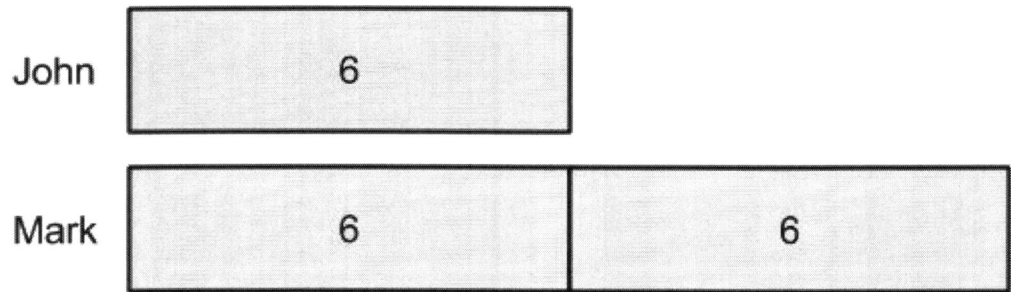

Number of balls Mark has = 2 x 6 = 12

Problem 3

John has 6 balls. Mark has 3 times as many balls as John. How many balls does Mark have?

Solution

Draw a box to represent the 6 balls John has and another three boxes of the same size to represent the number of balls Mark has.

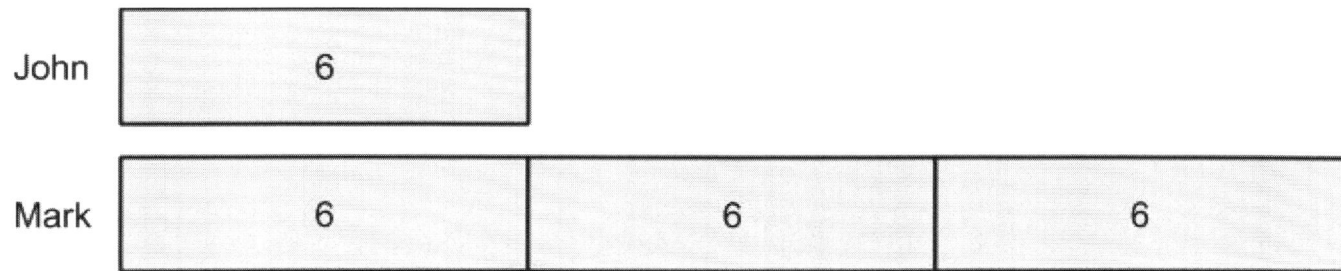

Number of balls Mark has = 3 x 6 = 18

Worksheet

1. Jim has 9 apples. Jenny has 8 apples more than Jim. How many apples does Jenny have?

2. Jim has 8 apples. Jenny has 4 times as many apples as Jim. How many apples does Jenny have?

3. Jim has 7 apples. Jenny has 5 times as many apples as Jim. How many apples does Jenny have?

4. Jim has 13 pears. Jenny has 6 pears less than Jim. How many pears does Jenny have?

Chapter 2. Using model diagrams to solve problems involving total

Problem 1

Ben has 6 boxes. Ben and William have 17 boxes altogether. How many boxes does William have?

Solution

Draw a box to represent the 6 boxes Ben has and a bigger box to represent the number of boxes William has.

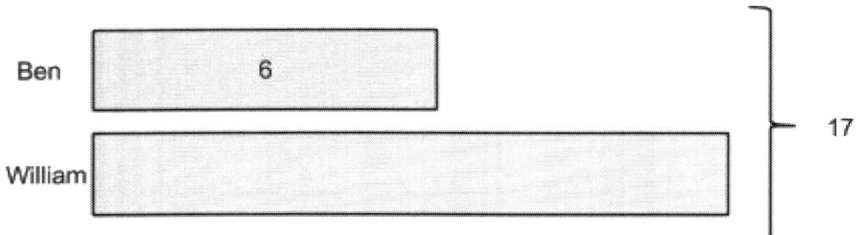

Number of balls William has = 17 - 6 = 11

Problem 2
Benn and William have 19 boxes altogether. William has 5 boxes more than Ben. How many boxes does William have?

Solution
Draw a box to represent the number of boxes Ben has and a bigger box to represent the number of boxes William has.

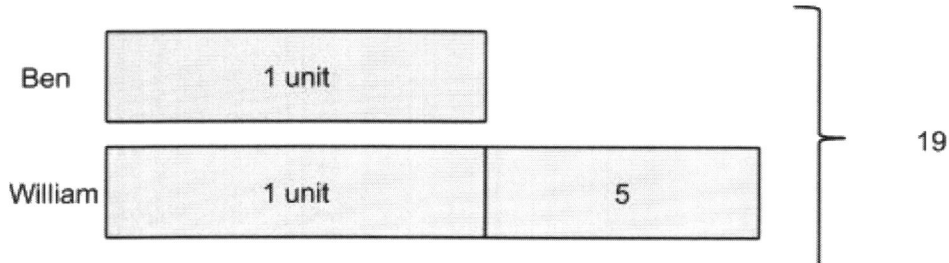

2 units = 19 - 5 = 14
1 unit = 14 ÷ 2 = 7

Number of boxes William has
= 1 unit + 5
= 7 + 5
= 12

Problem 3
William has twice as many apples as Ben. They have 24 apples. How many apples does William have?

Solution

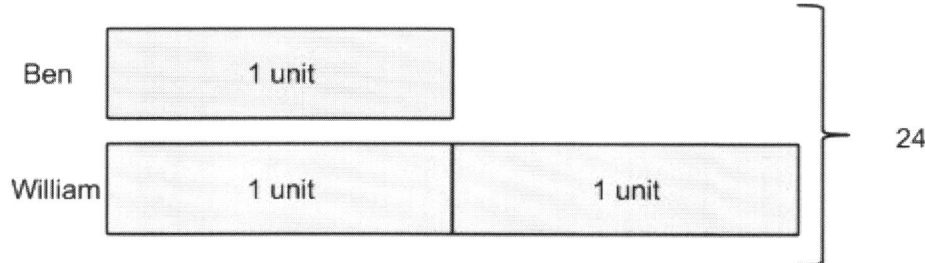

3 units = 24
1 unit = 24 ÷ 3 = 8

Number of boxes William has
 = 2 units
 = 2 x 8
 = 16

Problem 4
William has 2 times more apples than Ben. They have 24 apples altogether. How many apples does William have?

Solution

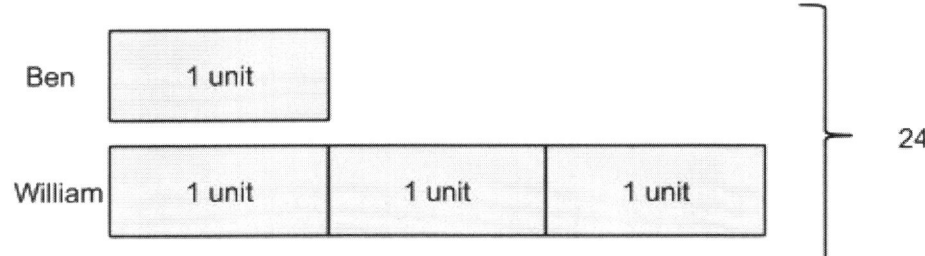

4 units = 24
1 unit = 24 ÷ 4 = 6

Number of boxes William has
 = 3 units
 = 3 x 6
 = 18

Worksheet 2

1. Jim has 5 apples. Jim and Jenny have 14 apples together. How many apples does Jenny have?

2. Jim and Jenny have 16 apples together. Jim has 4 more apples than Jenny. How many apples does Jenny have?

3. Jim has three times more apples than Jenny. They have 35 apples together. How many apples does Jim have?

4. The total weight of a dog and a mouse is 32 kg. The dog is 7 times as heavy as the mouse. Find the weight of the dog.

5. Debbie paid a total of $19 for a plate of spaghetti and a plate of fried rice. If the spaghetti costs $5 more than the fried rice, how much did the spaghetti cost?

Chapter 3. Using model diagrams to solve problems involving one party giving to another

Problem 1

Adam and Zach had the same number of books. Pete then gave Zach 9 books. Zach now has 4 times more books than Adam. How many books does Adam have?

Solution

Draw two boxes equal in size to represent that Adam and Zach have the same number of books.

After Pete gives 9 books to Zach, Zach has 4 times more books than Adam.

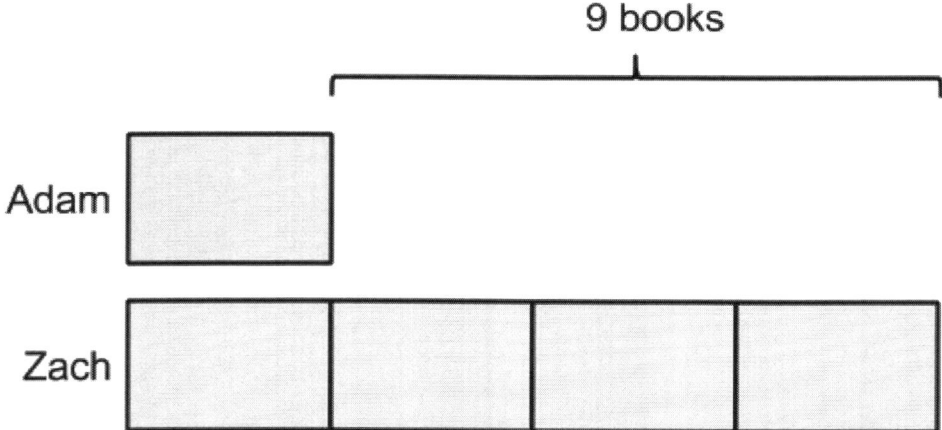

3 units = 9
1 unit = 9 ÷ 3 = 3

Number of books Zach has
= 4 units
= 4 x 3
= 12 books

Problem 2

Adam had two times as many books as Zach. After Adam gave 20 books to Peter, Zach now has 3 times as many books as Adam. How many books does Adam have at first?

Solution

At first, Adam had two times as many books as Zach

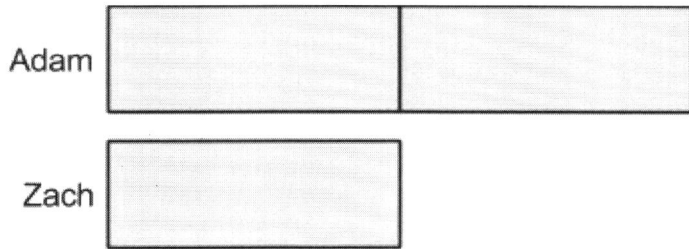

After Adam gave 20 books to Peter, Zach now has 3 times as many books as Adam

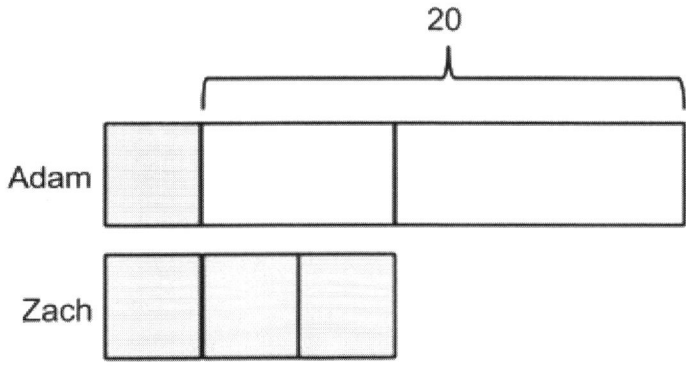

Cut Adam's boxes into units of the same size.

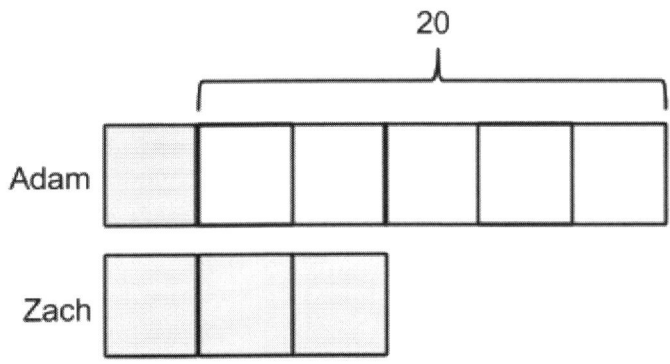

5 units = 20
1 unit = 20 ÷ 5 = 4
Number of books Adam had at first = 6 units = 6 x 4 = 24 books

12

Worksheet 3

1. Jim and Jenny have the same number of apples. After Jim gives Adam 18 apples, Jenny has 4 times as many apples as Jim. How many marbles does Jenny have?

2. Jeremy has twice as many sweets as Jack. After Jeremy gives Adam 21 sweets, Jack now has twice as many sweets as Jack. How many sweets does Jack have?

3. Jill has twice as much money as Joseph. After she spends $15, Joseph now has three times as much money as she. How much money did Joseph have?

4. Jasper have some cats and 50 dogs. After he sells away 18 dogs, he has four times as many dogs as cats. How many cats did he have?

5. Albert have some apples and oranges. He has 15 more apples than oranges. After he gave away 21 apples, he has three times as many oranges as apples. How many oranges does Albert have?

Chapter 4. Using model diagrams to solve problems involving one party receiving from another

Problem 1

Adam had twice as many books as Zach. After Zach received 21 books from Pete, Zach had twice as many books as Adam. How many books did Adam have?

Solution

Adam had twice as many books as Zach.

After Zach received 21 books from Pete, Zach had twice as many books as Adam.

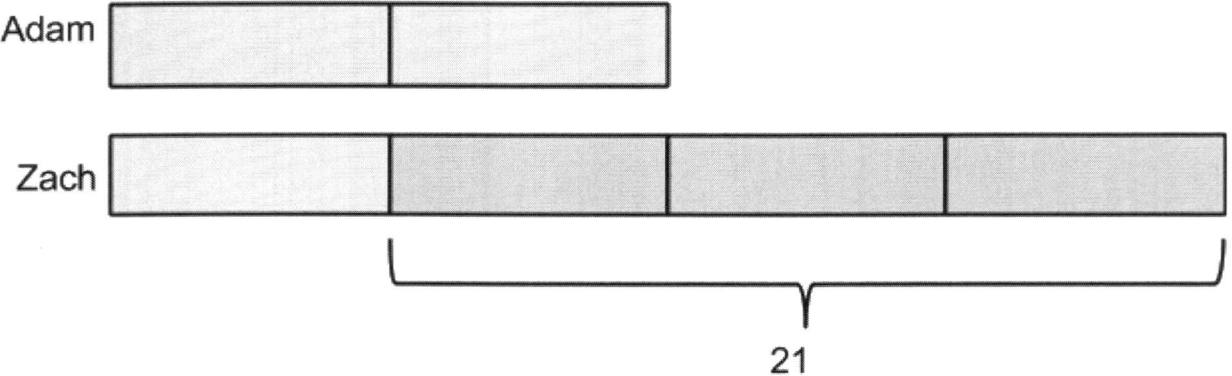

How many books did Adam have?
3 units = 21
1 unit = 21 ÷ 3 = 7

Adam had
2 units = 7 x 2 = 14 books

Problem 2

Adam and Zach had 50 sweets altogether. After Zach received 5 sweets from Pete, Adam had four times as many sweets as Zach. How many books did Adam have?

Solution

Draw two boxes of rough sizes for Adam and Zach to represent the total amount of sweets they have.

After Zach received 5 sweets from Pete, Adam had four times many sweets as Zach.

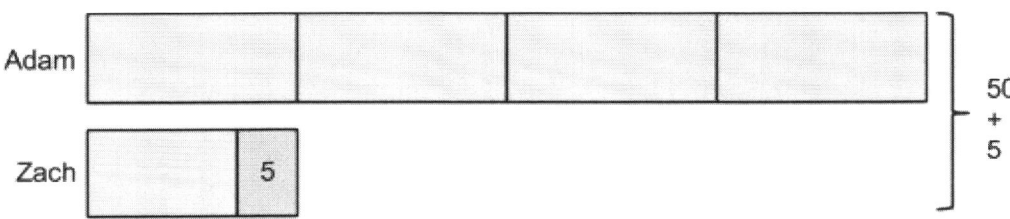

How many sweets did Adam have?

5 units = 50 + 5 = 55
1 unit = 55 ÷ 5 = 11

Number of sweets Adam has

4 units = 11 x 4 = 44 sweets

Worksheet 4

1. Jim has thrice as many apples as Jenny. After Jenny bought another 15 apples, she has twice as many apples as Jim. How many apples did Jim have?

2. Jim and Jenny have 50 oranges altogether. After Jenny bought another 6 oranges, Jim has 7 times as many oranges as Jenny. How many apples did Jim have?

3. Jim has 5 times as many pencils as Debby. After Debby bought another 12 pencils, Jim has thrice as many pencils as Debby. How many pencils did Jim have?

4. Ben had a total of 24 red and blue balls. After buying another 6 blue balls, he had an equal number of red and blue balls. How many red balls did he have?

5. Ben has 24 red balls and 10 blue balls. How many more blue balls must he buy so that he will have three times as many blue balls as red balls?

Chapter 5. Using model diagrams to solve problems involving both parties giving

Problem 1

Adam had 18 sweets. Zach had 14 sweet. After each ate the same number of sweets, Adam had twice as many sweets as Zach. How many sweets did they eat in total?

Solution

Adam had 18 sweets. Zach had 14 sweet.

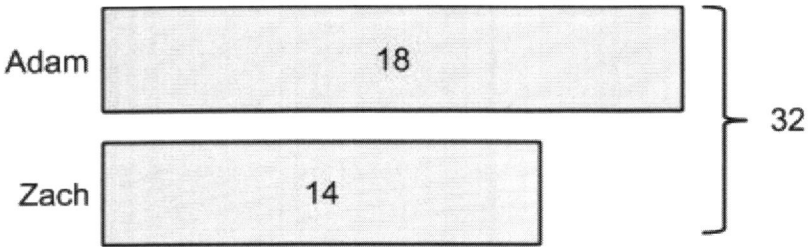

After each ate the same number of sweets, Adam had twice as many sweets as Zach.

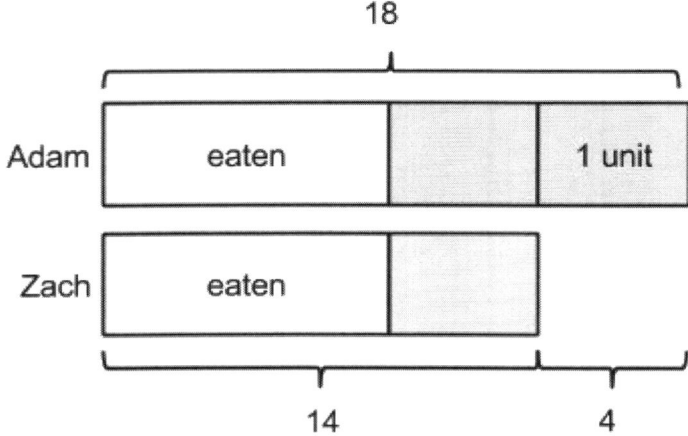

1 unit = 4 sweets
Zach had 4 sweets. Number of sweets he ate = 14 - 4 = 10 sweets.
Since both ate the same number of sweets, number of sweets eaten in total
= 10 x 2 = 20 sweets.

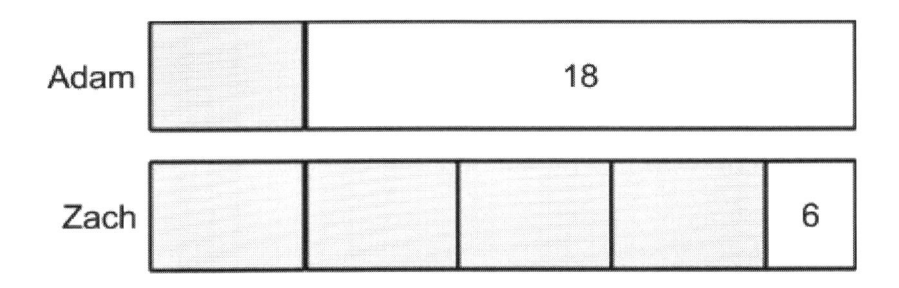

Problem 2

Adam and Zach had the same number of sweets. After Adam ate 18 sweets, and Zach ate 6 sweets, Zach had four times as many sweets as Adam. How many sweets did Adam have at first?

Solution

Adam and Zach had the same number of sweets.

After Adam ate 18 sweets, and Zach ate 6 sweets, Zach had four times as many sweets as Adam.

How many sweets did Adam have at first?

3 units = 18 - 6 = 12
1 unit = 12 ÷ 3 = 4

Number of sweets Adam had at first = 4 + 18 = 22

Worksheet 5

1. Jim had 30 sweets and Zach had 24 sweets. After each of them ate the same number of sweets, Jim had 3 times as many sweets as Zach. How many sweets did Jim ate?

2. Leon and Yvonne had the same number of books. Leon gave away 14 book and Yvonne gave away 6 book. Yvonne now had 5 times as many books as Leon. How many books did Leon have at first?

3. Leon is older than Yvonne by 15 years old. 18 years ago, Leon was twice as old as Yvonne. What is Leon's age now?

4. Ben had three times as many red balls than blue balls. After giving away 123 red balls and 21 blue balls, Ben has twice as many blue balls than red balls. How many blue balls did Ben have at first?

5. Tank A had 3 times as much water as tank B. Later, 12 litre of water was taken out from each tank. Tank A now had 6 times more water than tank B. How much water did tank A contain at first?

Chapter 6. Using model diagrams to solve problems involving both parties receiving

Problem 1

Adam had 18 sweets. Zach had 4 sweets. Each then bought an additional same number of sweets. Adam now had thrice as many sweets as Zach. How many additional sweets did each buy?

Solution

Adam had 18 sweets. Zach had 4 sweets.

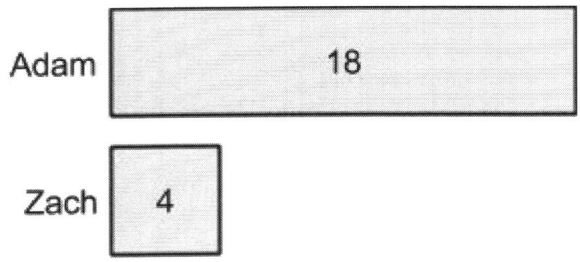

Each then bought the same number of sweets.

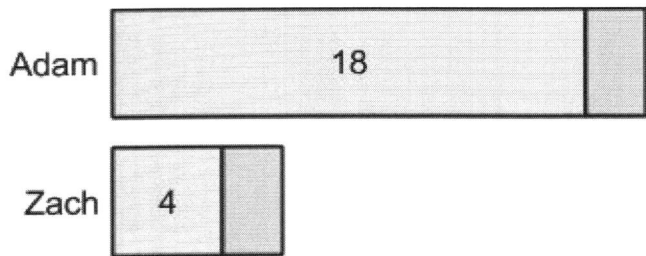

Adam now had thrice as many sweets as Zach. Adam's box can be re-drawn like the below to represent the number of additional sweets bought as 1 unit.

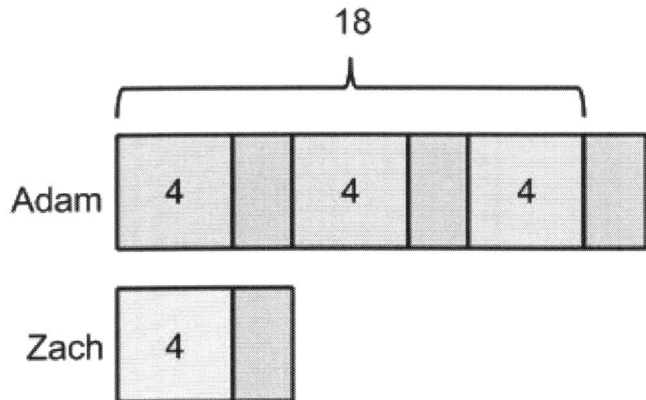

From the above diagram, 18 sweets is made up of

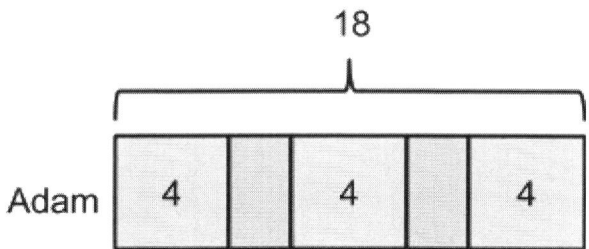

2 units of sweets will then be 18 - 4 - 4 - 4 = 6
1 unit = 6 ÷ 2 = 3

Number of additional sweets they each bought is 3 sweets.

Problem 2

Adam had 4 times as many sweets as Zach. Each then bought an additional 6 sweets. Adam now had 3 times as many sweets as Zach. How many sweets did Adam have at first?

Solution

Adam had 4 times as many sweets as Zach.

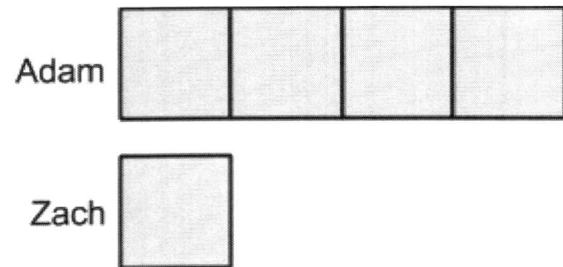

Each then bought an additional 6 sweets. Adam now had 3 times as many sweets as Zach.

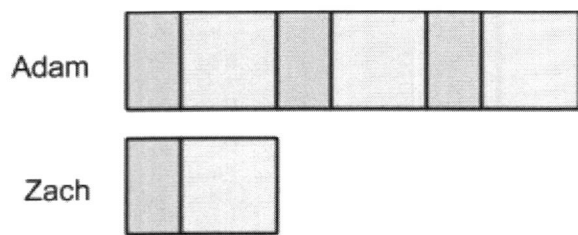

If we redraw Adam's box, we get the below two boxes.

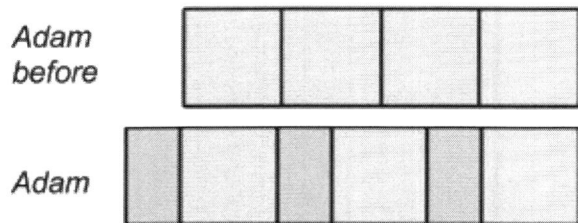

And we realize that four units is equal to 3 units + 6 + 6.

Therefore, 1 unit = 6 + 6 = 12

Zach had 1 unit at first.
Zach had 12 sweets at first.

25

Worksheet 6

1. Ben had 27 books. William had 3 books. After each additionally bought the same number of books, Ben now had 4 times as many books as William. How many additional books did each of them buy?

2. Angeline had 5 times as much money as Belle. Their parents then gave each of them another $4. Angeline now had 4 times as much money as Belle. How much money did Belle have at first?

3. Thaddeus is 4 times as old as Shannon. Thaddeus will be twice as old as Shannon in 12 years time. How old is Thaddeus now?

4. At a birthday party, there were twice as many cheeseburgers as hamburgers. A delivery man brought another 5 cheeseburgers and 34 hamburgers. There were now twice as many hamburgers as cheeseburgers. How many cheeseburgers were there at first?

5. Wendy had a total of 21 red and blue balls. After she bought another 10 red balls and 4 blue balls, she had 4 times as many red balls as blue balls. How many blue balls did she have at first?

Chapter 7. Using model diagrams to solve problems involving one party receiving and one party giving

Problem 1

Ben has twice as many books as Zach. After Ben buys 5 books and Zach gives away 2 books, Ben has thrice as many books as Zach. How many books does Ben have at first?

Solution

Ben has twice as many books as Zach.

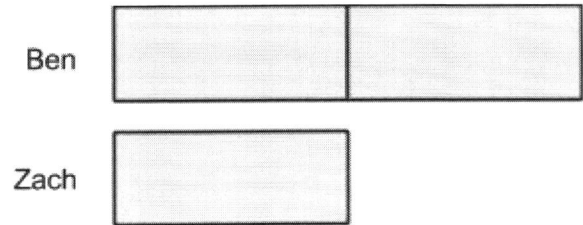

Ben buys 5 books and Zach gives away 2 books

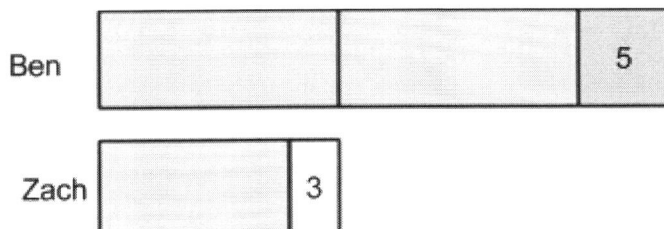

Ben has thrice as many books as Zach

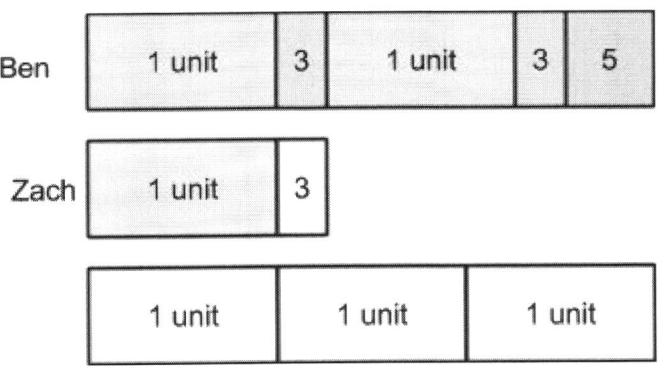

Therefore, 1 unit = 3 + 3 + 5 = 11

At first, Ben has 2 units + 3 + 3 = 22 + 6 = 28 books

Problem 2

Ben and Zach have 32 books together. After Ben gives away 5 books and Zach buys 3 books, Ben and Zach have the same number of books. How many books did Ben have at first?

Solution

It would be useful to draw the diagram 'backwards'.

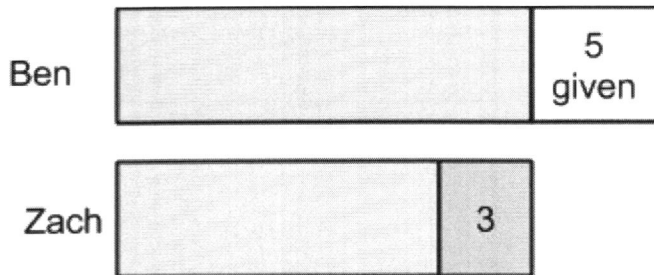

After Ben gives away 5 books and Zach buy 2 books, they should have left 32 - 5 + 3 = 30 books.

Each would then have 30 ÷ 2 = 15 books

Ben would have 15 + 5 = 20 books at first

Worksheet 7

1. Jason has 3 times as many sweets as William. After Jason buys 7 more sweets and William eats 3 sweets, Jason has 5 times as many sweets as William. How many sweets did Jason have at first?

2. Jeremy and Yvonne have $95 together. After Jeremy used up $24 and Yvonne received $33 from her father, both of them have the same amount of money. How much did Yvonne have at first?

3. Adam and Bob have the same number of toy cars. Adam then bought an additional same number of cars as what he had at first. Bob gave away 12 cars. Adam had 6 times as many cars as Bob. How many cars did Adam have at first?

4. Bobby has twice the amount of money that Cheryl has. Bobby then spent $12 and Cheryl earned some money such that she has 4 times the amount of money she had at first. If Cheryl now has 8 times as much money as Bobby, how much does Bobby have at first?

Chapter 8. Using model diagrams to solve problems involving one party giving to the other party

Problem 1

Ben has 18 books. Zach has 2 books. After Ben gives some books to Zach, Ben has thrice as many books as Zach. How many books did Ben give to Zach?

Solution

This sort of problem is simplified because we know that the total number of books does not change.

So total number of books = 18 + 2 = 20

If in the end, Ben has thrice as many books as Zach, then we have,

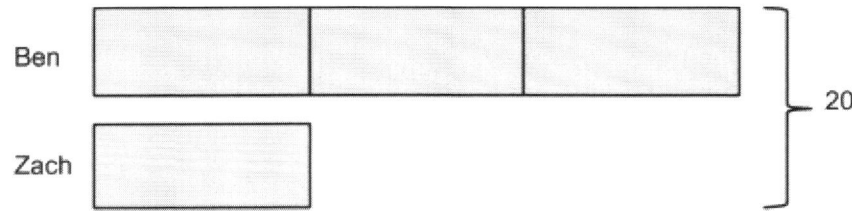

4 units = 20
1 unit = 20 ÷ 4 = 5

Zach has 5 books now. Thus, Ben gave Zach 5 - 2 = 3 books.

Problem 2

Ben has four times as many books as Zach. After Ben gives 6 books to Zach, both have the same number of books. How many books did Ben have at first?

Solution

After Ben gives 6 books to Zach, both have the same number of books. Split the 3rd unit into halve.

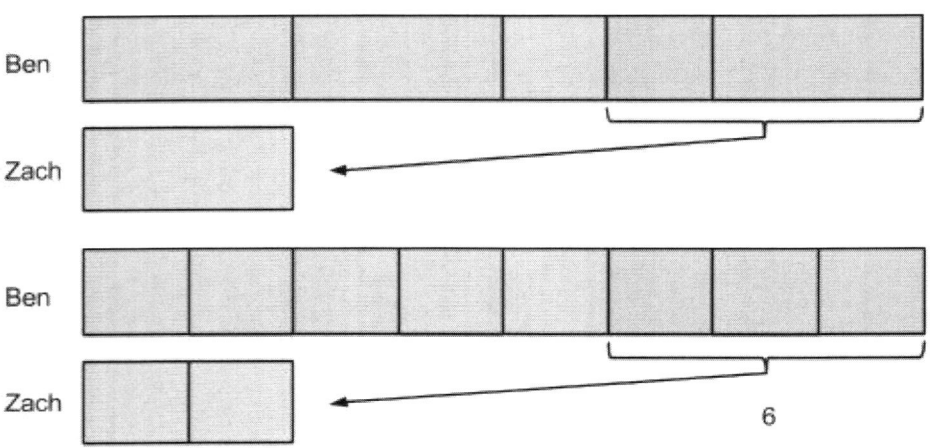

So, 3 units = 6
1 unit = 6 ÷ 3 = 2

Number of books Ben have at first is 8 units
= 8 x 2
= 16 books

Worksheet 8

1. Jason has 41 sweets. William has 3 sweets. After Jason gives some sweets to William, Jason has 3 times as many sweets as William. How many sweets did Jason give to William?

2. Jason and William have 54 sweets together. After Jason gives 4 sweets to William, Jason has 5 times as many sweets as William. How many sweets did William have at first?

3. Jason has four times as many balls than William. After Jason gives William 3 balls, Jason has three times as many balls as William. How many balls did William have at first?

4. Alex and Barbara have a total of 42 pencils. After Alex gave Barbara 3 pencils, Alex still has 4 more pencils than Barbara. How many pencils did Alex have at first?

5. There are both red balls and blue balls in a box. The number of red balls is 4 times as many blue balls. After painting 8 red balls into blue balls, there were twice as many red balls as blue balls. How many blue balls were there at first?

Chapter 9. Using model diagrams to solve problems involving insufficient money

Problem 1

Ben wanted to buy 8 toy cars but he was short of $6. He bought 3 toy cars instead and had $19 left. How many money did he have at first?

Solution

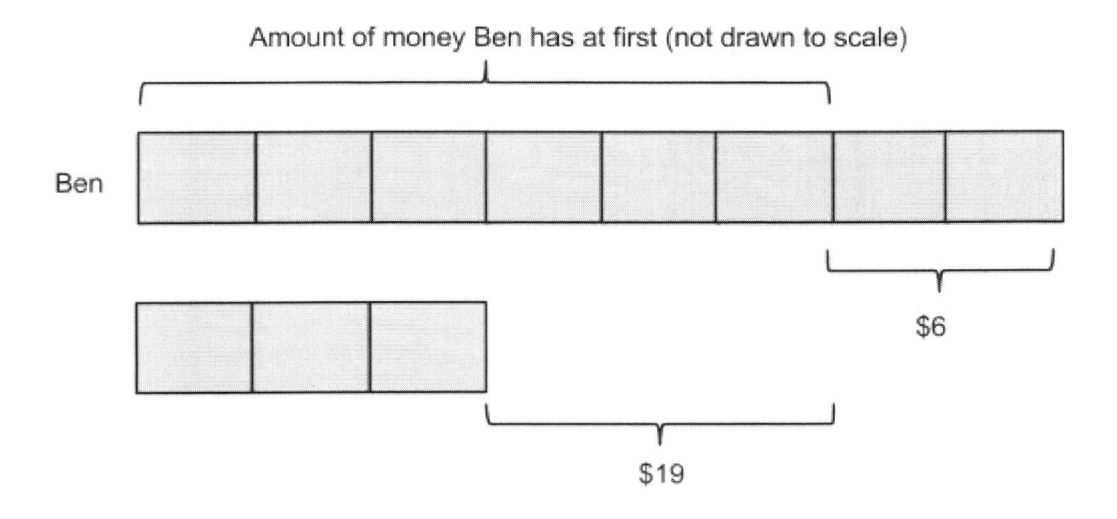

Amount of money Ben has at first (not drawn to scale)

Ben

$6

$19

Therefore, 5 toy cars will cost $6 + $19 = $25.
5 units = $25
1 unit = $25 ÷ 5 = $5

There are two ways of calculating how much he had at first.

Either, considering that he wanted to buy 8 toy cars but he was short of $6
8 units - $6
= 8 x 5 - 6
= $34

OR he bought 3 toy cars instead and had $19 left.

3 units + $19
= 3 x 5 + 19
= $34

Worksheet 9

1. Jason wanted to buy 7 cards but was short of $6. He bought 3 cards and had $2 left. How much money did he have at first?

2. Claire wanted to buy 3 pens but was short of $10. She bought 2 pens instead and had $7 left. How much money did she have at first?

3. Jerry wanted to buy some toy guns. If he were to buy 6 toy guns, he would have $5 left. If he were to buy 2 toy guns, he would have left $29. How much did a toy gun cost?

4. Jerry wanted to buy some toy guns. If he were to buy 7 toy guns, he would have $8 left. If he were to buy 3 toy guns, he would have left $32. How much did a toy gun cost?

Chapter 10. Using model diagrams to solve problems involving buying objects that are multiples of another

Problem 1

Jack bought 4 red balls and 4 blue balls for $72. Each blue ball cost twice as much a red ball. How much did a red ball cost?

Solution

Jack bought 4 red balls and 4 blue balls for $72.

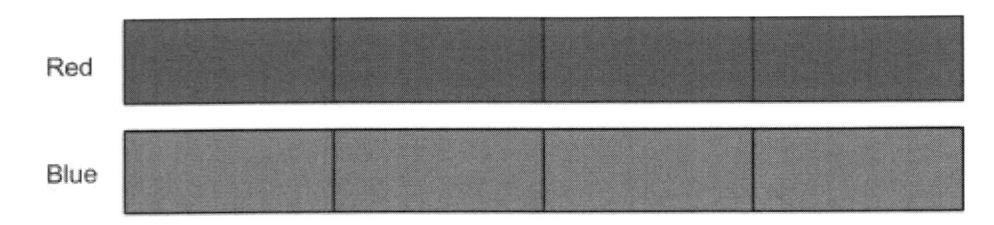

To represent that each blue ball cost twice as much a red ball, we draw twice the number of units for blue balls.

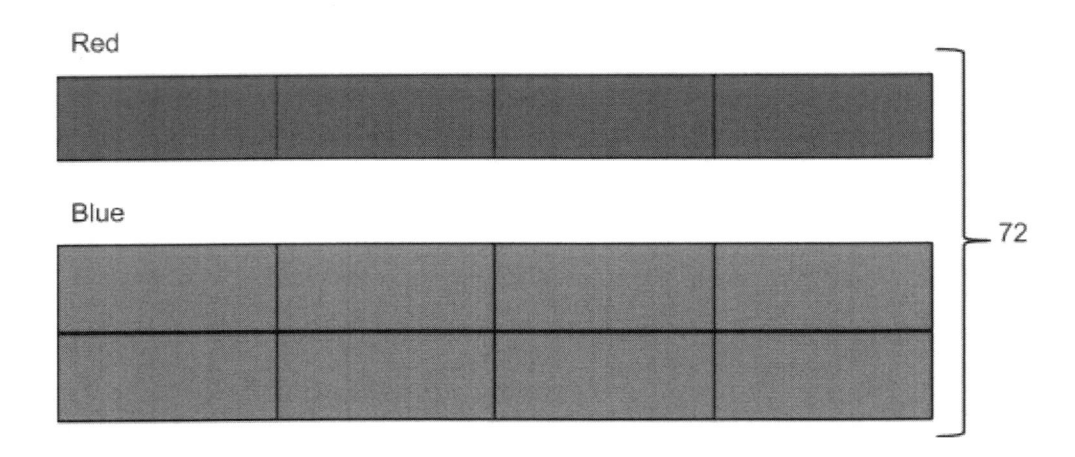

12 units = $72
1 unit = 72 ÷ 12 = $6

Each red ball cost $6.

Problem 2

Jack bought 3 red balls and 5 blue balls for $84. Each blue ball cost $4 more than a red ball. How much did a red ball cost?

Solution

Jack bought 3 red balls and 5 blue balls for $84.

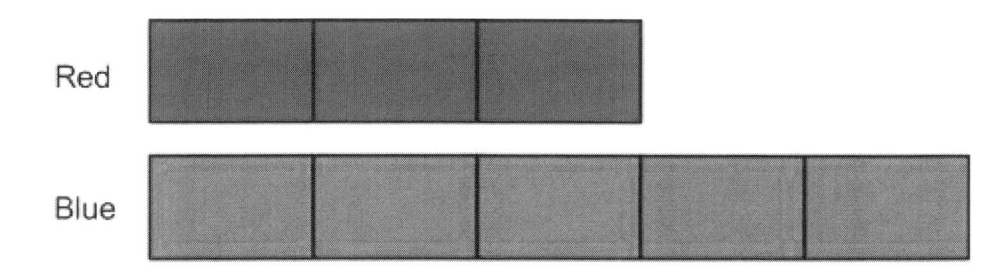

Each blue ball cost $4 more than a red ball. Add a box of $4 to each blue unit to represent the cost for blue balls.

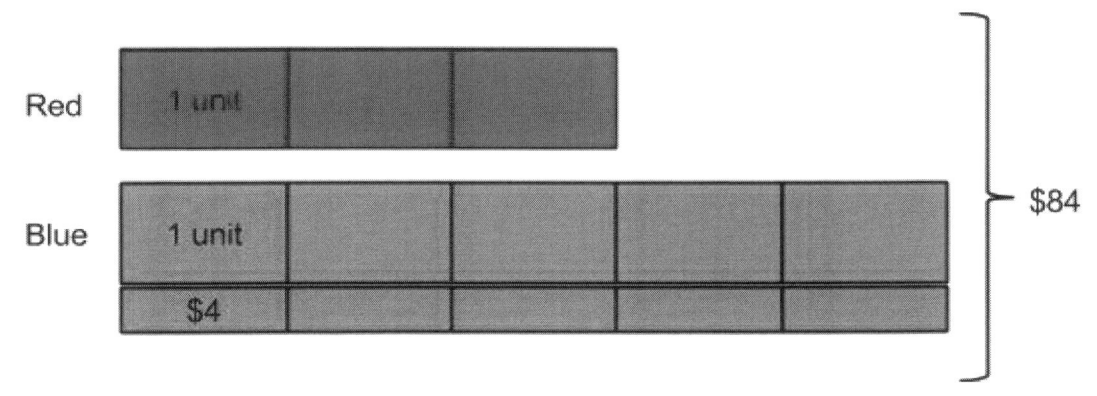

In total, we have 8 units and $4 x 5 = $20

8 units + $20 = $84
8 units = $84 - $20 = $64
1 unit = $64 ÷ 8 = $8

Each red ball cost $8.

Worksheet 10

1. The total cost of 4 toy cars and 3 toy rockets is $78. Each toy rocket costs as much as 3 toy cars. Find the cost of a toy car.

2. May had $160 worth of $1 and $2 dollar notes. She has 3 times as many $1 notes as $2 notes. How many notes does she have altogether?

3. Jack has 4 times as many 10 cent coins as 50 cent coins. The total value of his coins is $18. How many 50 cents does he have?

4. Finn bought 3 red balls and 4 blue balls for $54. 3 red balls cost the same as 2 blue balls. How much did each red ball cost?

Chapter 11. Using model diagrams to solve problems where the prices of individual objects are unknown

Problem 1

How much does a toy bike cost if,

🚗 + 🚗 + 🚗 + 🏍 + 🏍 = $58

🚗 + 🚗 + 🏍 + 🏍 = $48

Solution

Represent the objects in a model diagram

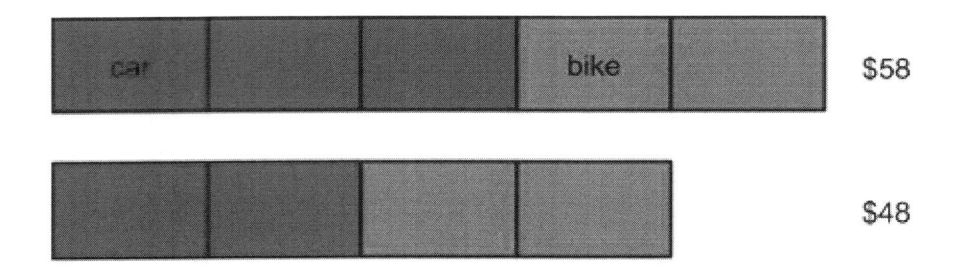

| car | | | bike | | $58

| | | | | $48

If we take the top bar and subtract the bottom bar away from it, we get:
Cost of a toy car = $58 - $48 = $10
Cost of 2 bikes = $48 - $10 - $10 = $28
Cost of 1 bike = $28 ÷ 2 = $14

Problem 2

How much does a toy bike cost if,

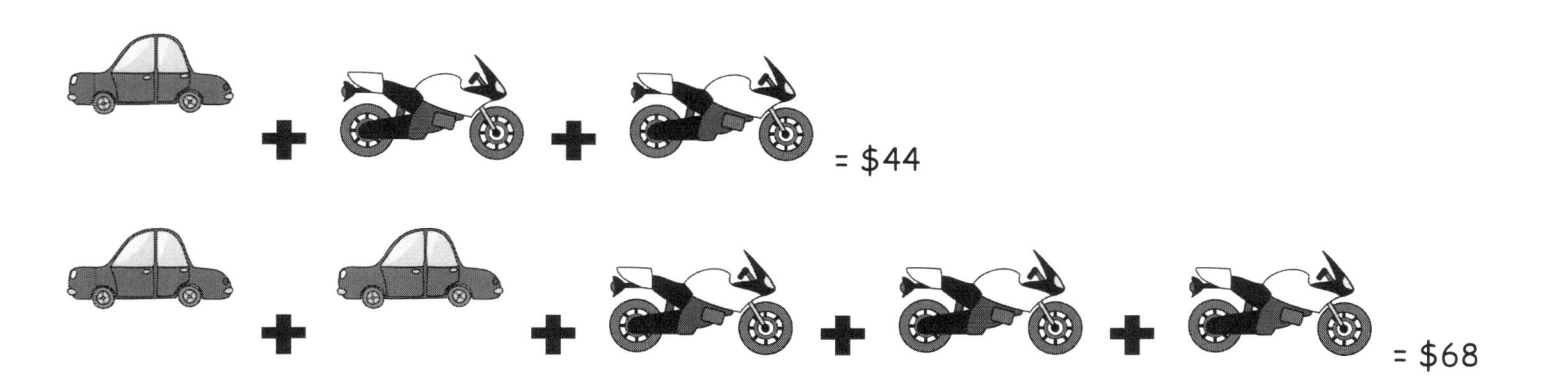

car + bike + bike = $44

car + car + bike + bike + bike = $68

Solution

Represent the objects in a model diagram

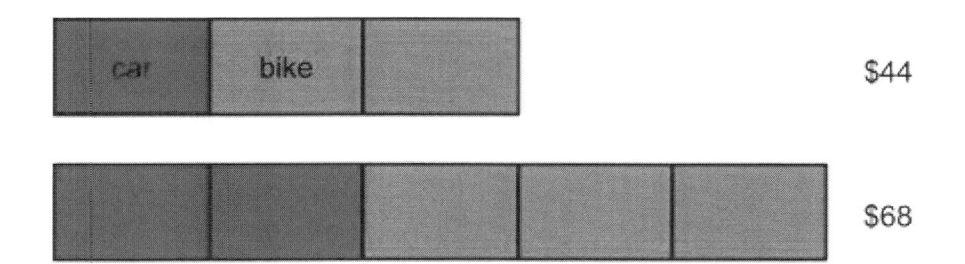

| car | bike | | $44

| | | | | | $68

The trick is to eliminate away one of the objects. We do by multiplying the first bar by 2

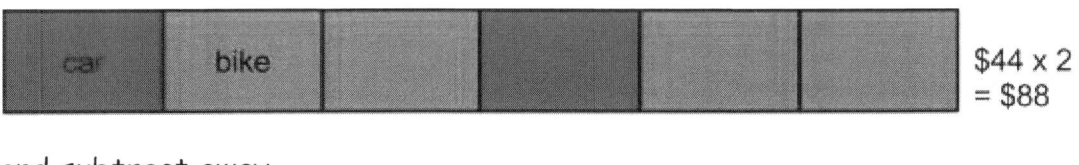

| car | bike | | | | | $44 x 2 = $88

and subtract away

| | | | | | $68

Thus, cost of 1 bike = $88 - $68 = $20

Worksheet 11

1. How much does a toy bike cost?

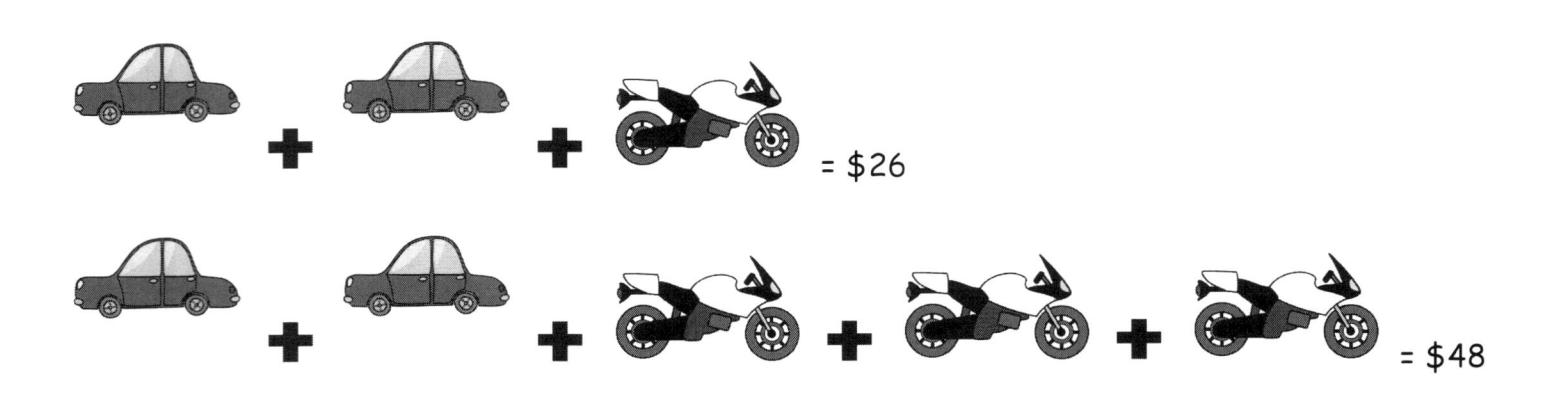

   car + car + bike = $26

   car + car + bike + bike + bike = $48

2. How much does a toy bike cost?

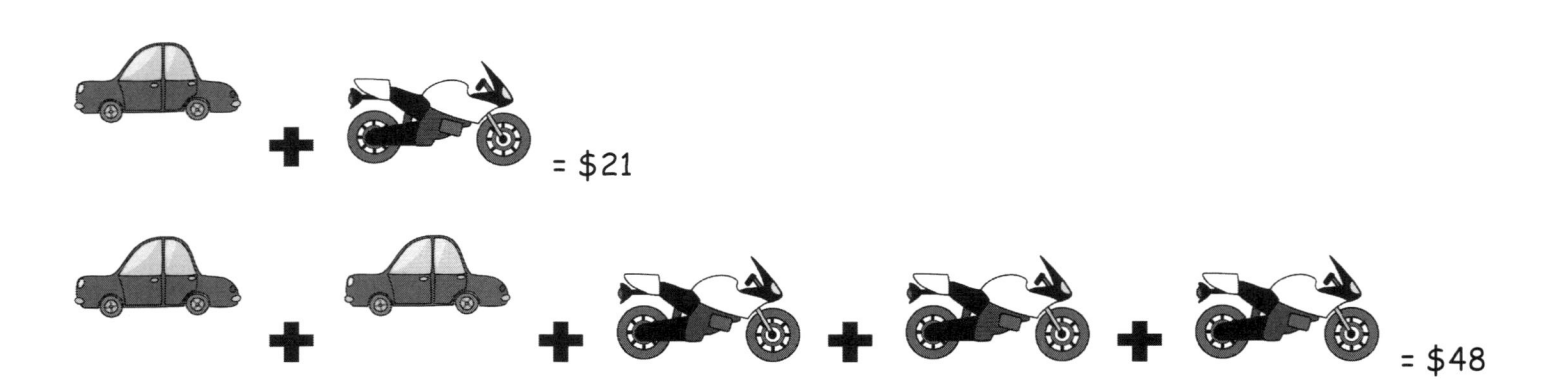

   car + bike = $21

   car + car + bike + bike + bike = $48

3. How much does a toy bike cost?

🚗 + 🚗 + 🏍 + 🏍 + 🏍 = $21

🚗 + 🏍 + 🏍 + 🏍 + 🏍 = $18

4. Three apples and 2 pears cost $46. 2 apples and 4 pears cost $50. How much does a pear cost?

5. Five apples and 3 pears cost $60. 2 apples and 2 pears cost $28. How much does an apple cost?

Chapter 12. Using model diagrams to solve 'filling up tank' problems

Problem 1

Tap A can fill up an empty tank in 4 hours and Tap B can fill up the same empty tank in 2 hours. How long will it take to fill up the empty tank if both taps are turned on at the same time?

Solution

Use a 4 by 2 grid to represent the tank.

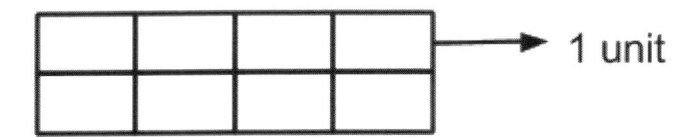

→ 1 unit

Tap A can fill up an empty tank in 4 hours. Amount of water Tap A fills in an hour:

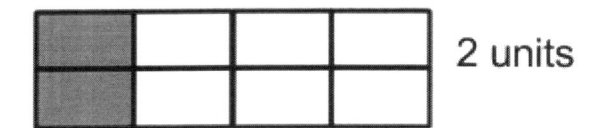

2 units

Tap B can fill up an empty tank in 2 hours. Amount of water Tap B fills in an hour:

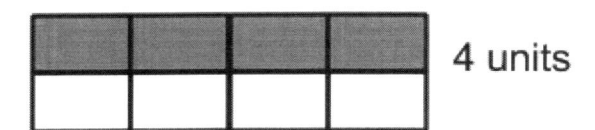

4 units

In an hour, both taps can fill 4 + 2 = 6 units

The full capacity of the tank is 8 units.

So, time taken to fill up the tank by both taps
= 8 ÷ 6
= $8/6$
= $4/3$
= $1\ 1/3$ hours

Worksheet 12

1. Tap A can fill up an empty tank in 3 hours and Tap B can fill up the same empty tank in 6 hours. How long will it take to fill up the same empty tank if both taps are turned on at the same time?

2. Tap A can fill up an empty tank in 7 hours and Tap B can fill up the same empty tank in 4 hours. How long will it take to fill up the same empty tank if both taps are turned on at the same time?

3. Tap A can fill up an empty tank in 15 hours. Tap B can fill up the same empty tank in 5 hours. Tap C can fill up the same empty tank in 3 hours. How long will it take to fill up the same empty tank if all three taps are turned on at the same time?

4. Six children can eat a cake in 1 hour. How long will it take for 5 children to eat the same cake?

Chapter 13. Using model diagrams to solve fraction problems

Problem 1

Bob has 52 red and blue balls. Of the 52 balls, 12 are rubber balls and the rest are plastic balls. $4/5$ of the red balls and $2/3$ of the blue balls are made of plastic. How many red balls does he have?

Solution

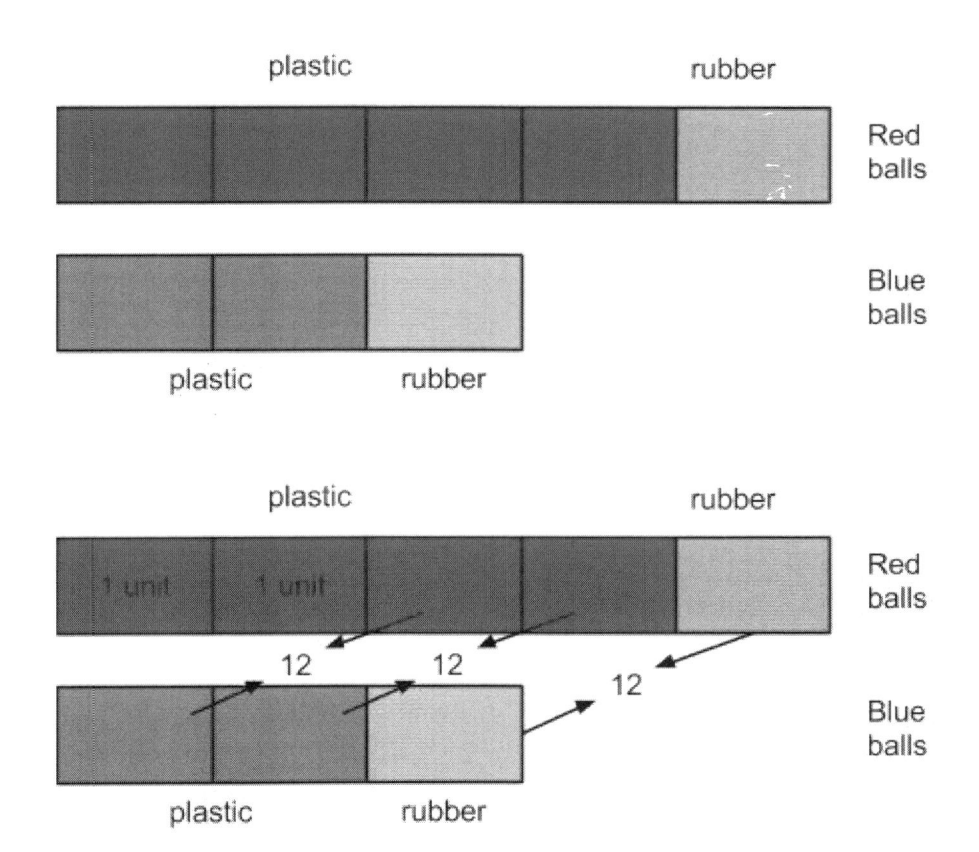

We can form 3 sets of 12.
3 x 12 = 36
The remaining 2 units will be 52 - 36 = 16
1 unit = 16 ÷ 2 = 8

Number of red balls = 5 units = 5 x 8 = 40 balls

Worksheet 13

1. Sharon has 42 red and blue balls. 8 of them are made of plastic. $5/6$ of the red balls and $4/5$ of the blue balls are made of rubber. How many red balls does she have?

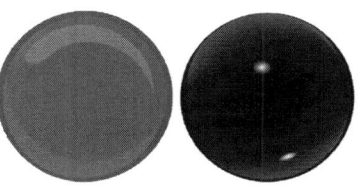

2. 30 students went for an excursion. 8 of them wore hats. $2/9$ of the girls and $1/3$ of the boys wore hats. How many girls went for the excursion?

3. A room had 60 red and white seats. 8 of the seats were reserved and the rest were not. $1/8$ of the red seats and $1/4$ of the white seats were reserved. How many red seats were in the room?

4. There were 95 cars and lorries in a parking lot and 20 of them ran on diesel. The rest ran on petrol. $7/8$ of the cars and $2/3$ of the lorries ran on petrol. How many cars were in the parking lot?

Answers

Worksheet 1: 1) 17   2) 32   3) 35   4) 7
Worksheet 2: 1) 9   2) 6   3) 28   4) 28   5) 12
Worksheet 3: 1) 24   2) 14   3) 9   4) 8   5) 9
Worksheet 4: 1) 9   2) 49   3) 90   4) 15   5) 52
Worksheet 5: 1) 21   2) 16   3) 48   4) 45   5) 60
Worksheet 6: 1) 5   2) 12   3) 24   4) 24   5) 3
Worksheet 7: 1) 33   2) 19   3) 18   4) 16
Worksheet 8: 1) 8   2) 5   3) 9   4) 26   5) 12
Worksheet 9: 1) 8   2) 41   3) 6   4) 6
Worksheet 10: 1) 6   2) 128   3) 20   4) 6
Worksheet 11: 1) 11   2) 6   3) 3   4) 6   5) 9
Worksheet 12: 1) 2   2) $2\frac{6}{11}$ 3) $\frac{3}{5}$ 4) $\frac{6}{5}$
Worksheet 13: 1) 12   2) 18   3) 56   4) 56

Full hand worked out answers can be obtained by sending a request to support@i-ducate.com

Printed in Great Britain
by Amazon